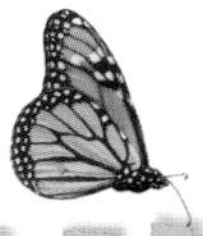

Animals

written by Lucy Floyd

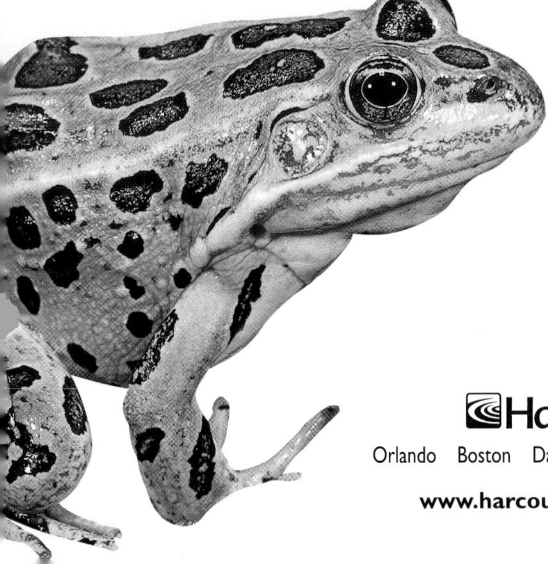

Harcourt

Orlando Boston Dallas Chicago San Diego

www.harcourtschool.com

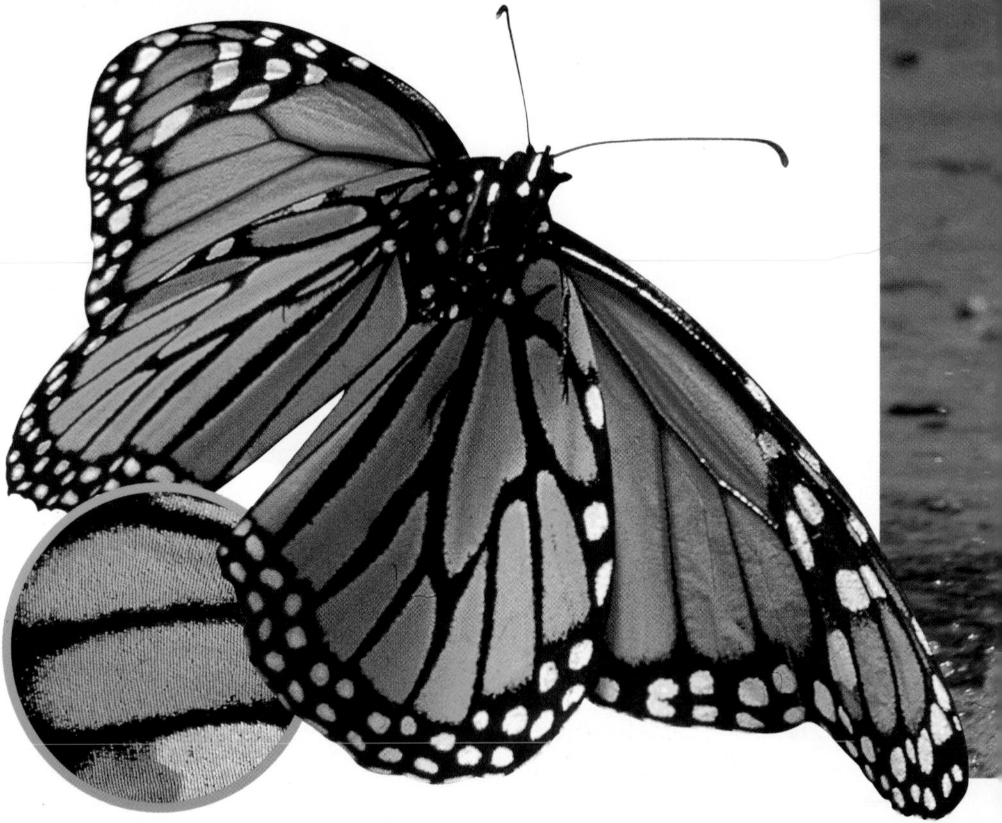

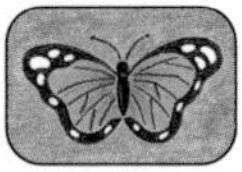

a butterfly

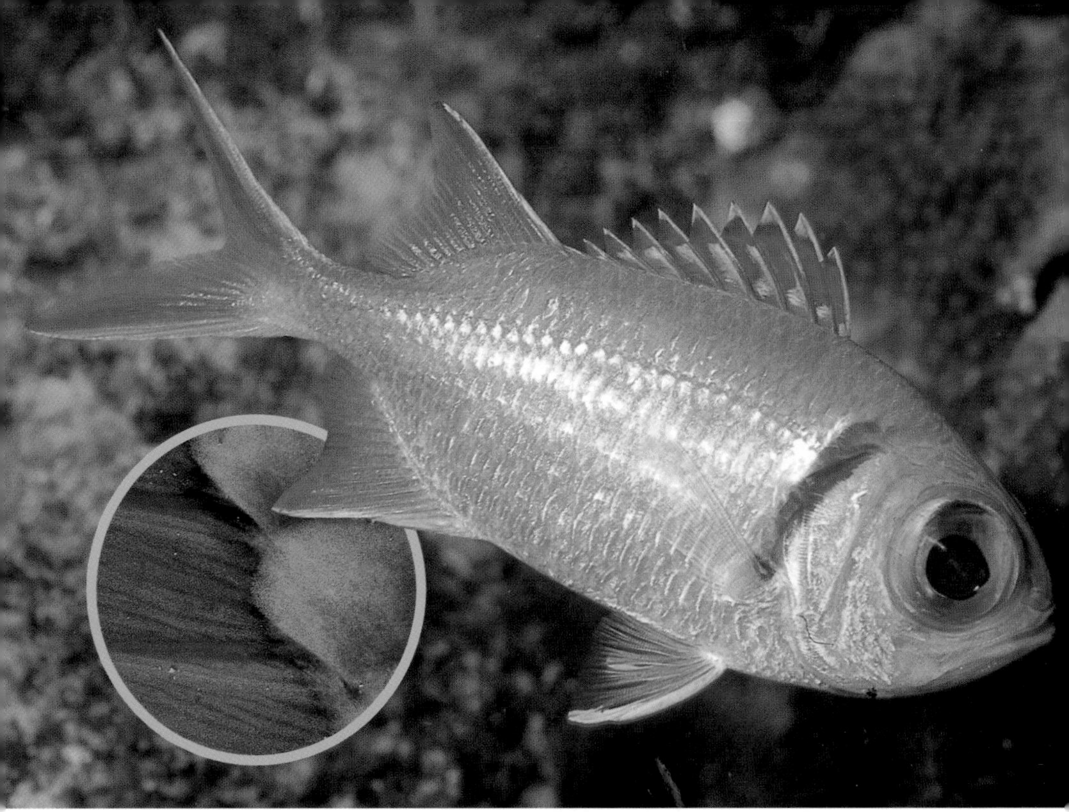

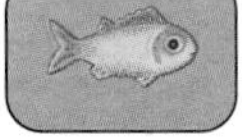

a fish

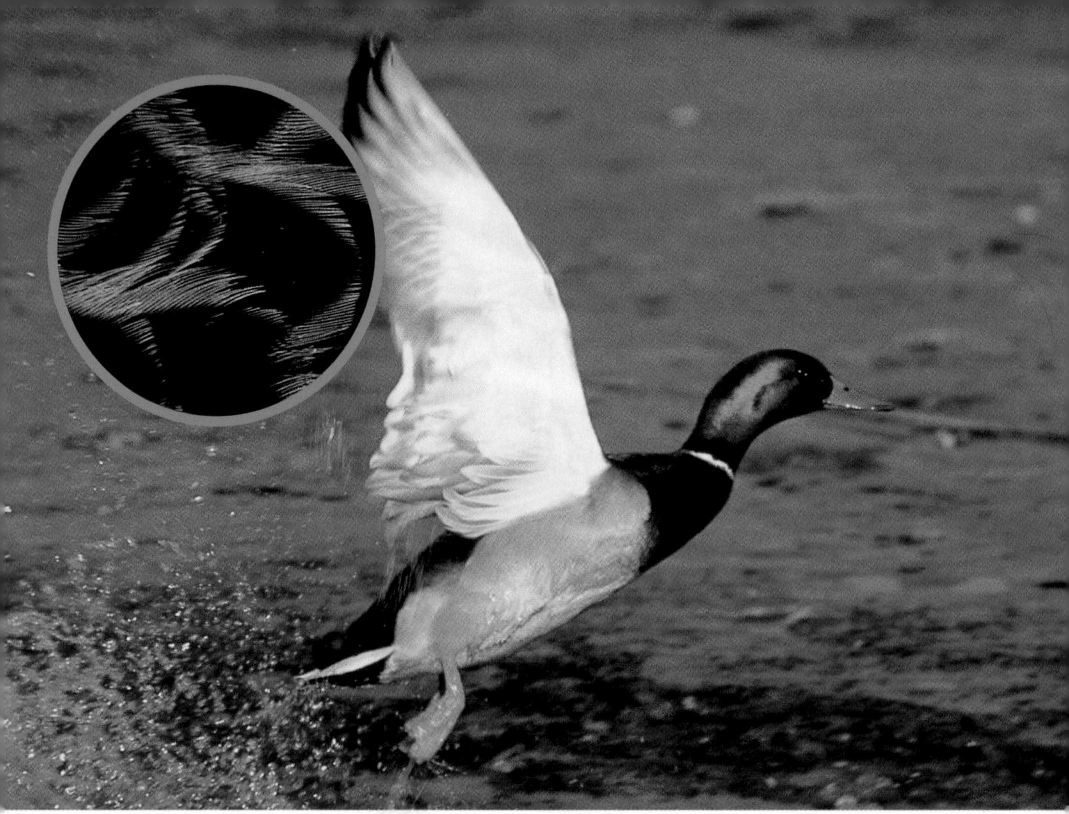

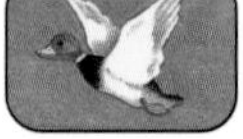

a duck

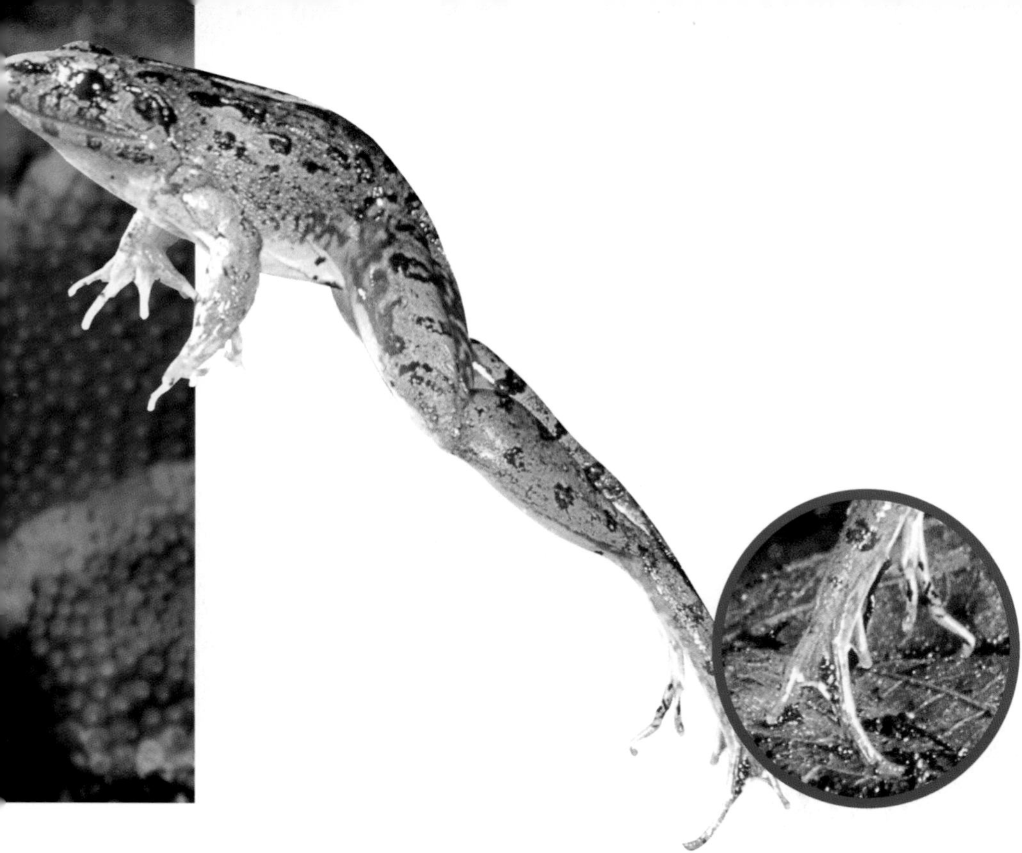

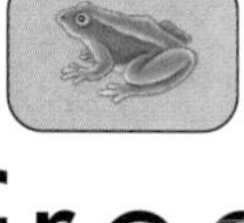

a frog

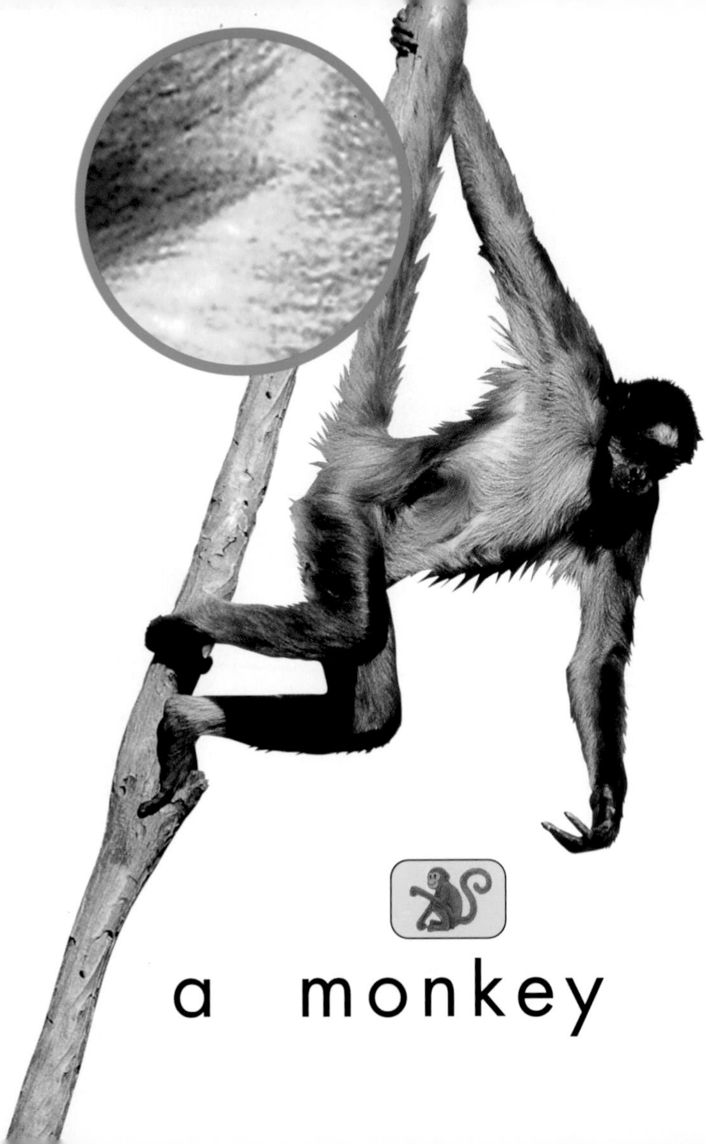

a monkey

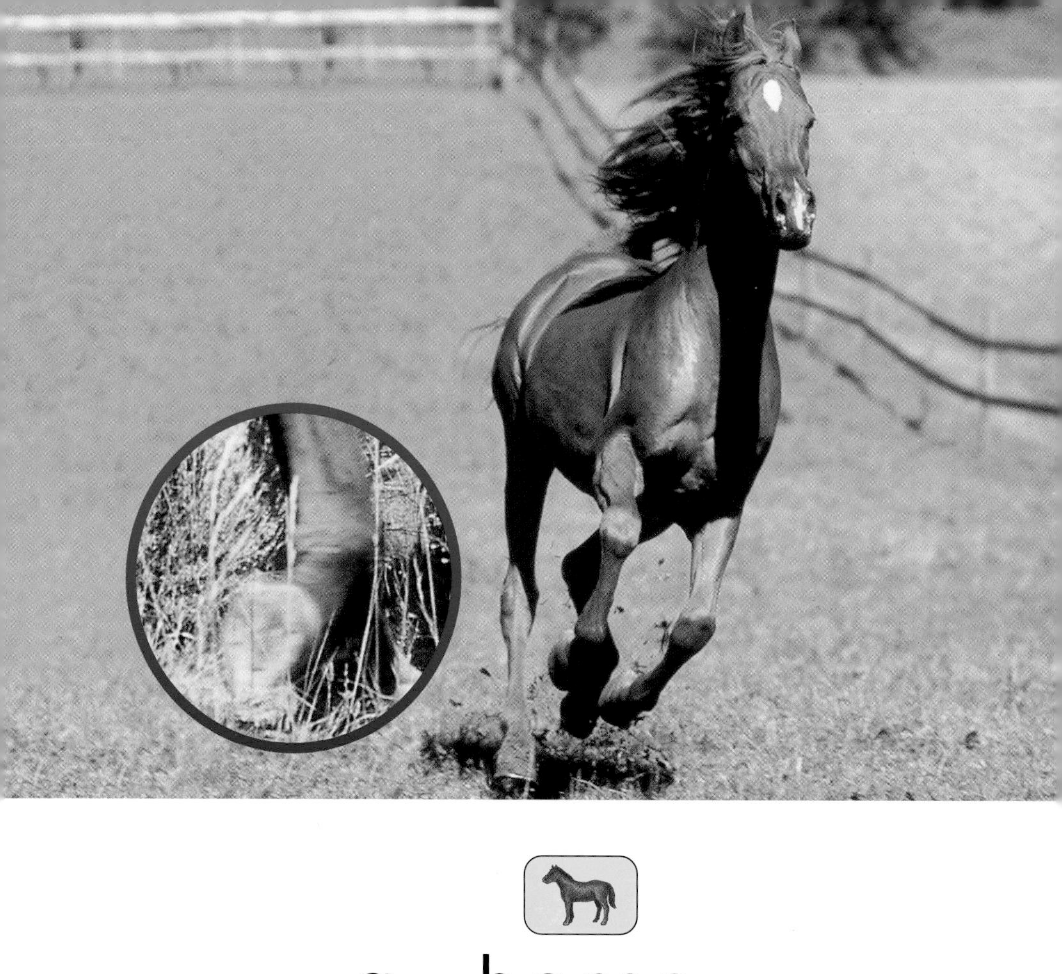

a horse

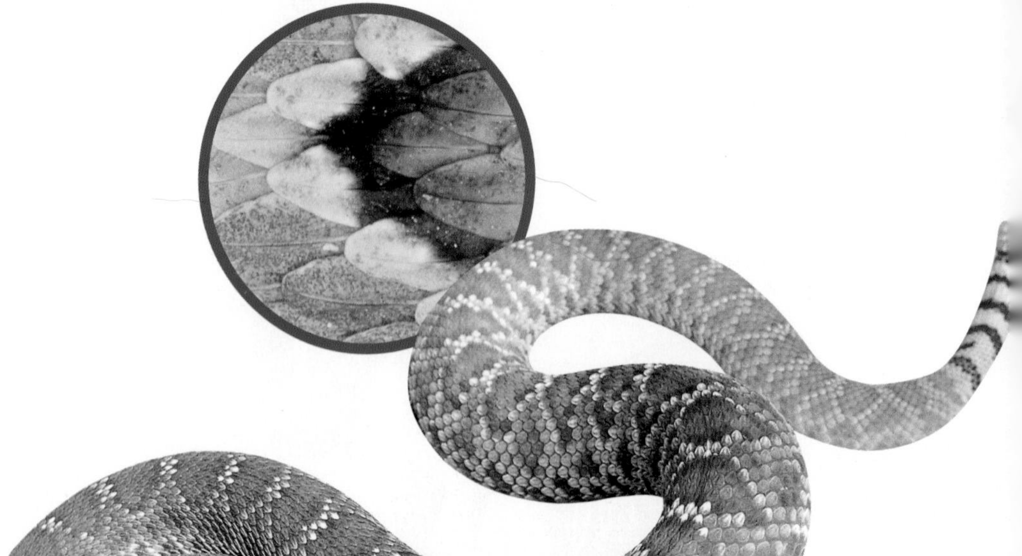

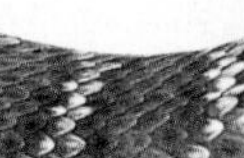

a snake